HYPNOSIS 101

Basic Hypnotic Techniques

A COMPLETE INTRODUCTION TO HYPNOSIS

WITH BASIC INDUCTION TECHNIQUES

MONOLOGUES AND DEEPENING

PROCEDURES

BY

Larry M. McDaniel

© June, 2017 by Larry M. McDaniel
All Rights Reserved
ISBN: 978-1-304-18372-9

FOREWORD

It is the purpose of this course to offer the serious student of hypnosis the basic fundamentals of the art. This course represents methods and techniques, which are outlined in a concise and clear presentation. It is intended to be usable for the student and instructor alike for basic training in hypnosis.

This course doesn't cover hypnotherapy, hypnoanalysis or any involved and complex therapy but is designed to be the groundwork from which the student can spring into higher training and greater achievement. It is assumed that the student is genuinely interested and willing to devote study time and effort to the overall goal of being helpful to his fellow man.

This course is presented in an outline form for easy reference and as a teaching aid. Clear instructions as to how to go about hypnosis is given along with a general coverage of several facets of hypnosis and an explanation of each. It is prepared with the express desire to see students move forward with a spark of fire and enthusiasm for excellence in his or her study of the art and science of hypnotherapy.

BACKGROUND

The author, Larry M. McDaniel, holds a BA degree in Psychology from Oglethorpe University in Atlanta, Georgia. His post-graduate studies in Advanced Psychotherapeutic Analysis were done through the University of California in Los Angeles in a Continuing Education Program that was administered by William J. Bryan, Jr., M.D., Ph.D., LLD, JD, Director of the American Institute of Hypnosis in Los Angeles, California. Dr. Bryan also was head of the College of Medical Hypnosis during the 1960's until his death in the mid 1970's.

Mr. McDaniel was trained in all phases of medical hypnosis and was admitted to the Faculty of the American Institute of Hypnosis in 1969. He taught physicians and psychologists with the American Institute throughout the 1970's and continued on in his own practice at the American Institute for Human Development in Atlanta, Georgia after the death of Dr. Bryan.

Mr. McDaniel was certified by the State of Georgia Merit System as Clinical Psychologist in 1976. He was Founder and President of the National Hypnotherapy Association that was headquartered in the Atlanta area for many years.

TABLE OF CONTENTS

1. DEFINITION AND TERMINOLOGY..6

   Hypnosis - Post Hypnotic Suggestions - Auto Hypnosis- Waking Hypnosis - Conscious Mind - Subconscious Mind - Catalepsy - Somnambulism - General Relaxation - Eye Closure - Muscle Rigidity - Motor Movements - Partial Anesthesia - Total Anesthesia - Partial Response to Post Hypnotic Suggestions - Total Response to Hypnotic Suggestions - Positive Visual Hallucinations - Negative Visual Hallucinations - Partial Amnesia Upon Awakening - Total Amnesia Upon Awakening - Hypnotic Coma (Ultra-Depth Hypnosis)

2. PRELIMINARY TESTING..10

   Heavy and Light Arm Technique - Hand Clasp Test - Falling Backward Test - Progressive Relaxation Process - Swinging Pendulum Test.

3. EXTERNAL FACTORS..14

   Personality Clashes - Professionalism - Office Arrangement - Secretary and/or Receptionist - Telephone Voice - Hypnotic Techniques - Personal Induction - Tape Induction - Tailor the Technique to the Subject - Emotional Factors - An Upset Subject - Admiration of the Therapist - Diplomas - Other Factors - Cost Spousal Support or Opposition - Results.

4. INDUCTION TECHNIQUES..20

   Swinging Pendulum - Eye Fixation on a stationary Point - Counting and/or Pyramiding - Visualization o- Stroking - hand and Arm Levitation - Peaceful Regressive Techniques - Tension Relaxation Alternation - Elevator.

5. INDUCTION MONOLOGUE..25

   Hypnotic Induction - Self Hypnosis Induction.

6. ADDITIONAL FACTORS..34

   Awakening Methods - Simple Count - Long Count - Resistant Subjects - Direct Command Approach - Permissive Approach - Sleep Approach - When All Other Methods Fail - Walk Subject - Sleep It Off - Medical Assistance - Problem Solving Methods - TV Techniques - Delayed Answer Technique - Dwelling On Solution - Positive Suggestion Application - Self-Suggestion - Supplemental

Tapes - Mirror Techniques - Mechanical Aids - Pendulum - Light - Metronome
Revolving Spiral - Fixation Point - Music - Chairs - Automated Hypnosis
Dangers of Hypnosis - Removal of Suggestions - Confabulation - Hysteria
Physical Dangers - Heart Patients - Emotionally Disturbed Individuals - Other
Dangers - Transference - Malpractice - Producing Sudden Changes.

CONCLUSION..42

CHAPTER ONE

DEFINITIONS AND TERMINOLOGY

Before entering into a study of any subject, an understanding of the terminology used within the confines of that study must be understood. The terms listed herein will be used throughout this course with the anticipation that the student will understand them clearly and for that purpose they are defined and explained in the first segment.

Definitions other than those listed herein are available from different sources and may vary somewhat from those you will see here. Slight variances will not make any appreciable differences in your understanding of the course material.

HYPNOSIS: An induced state of concentration and hypersuggestability brought about in a subject by the subject's own means or by an external controller.

POST-HYPNOTIC SUGGESTIONS: Suggestions given to a subject while in the hypnotic state that will be carried out upon awakening.

AUTO-HYPNOSIS: Self-Hypnosis.

WAKING HYPNOSIS: Suggestions given to a subject when in the normal waking state that tend to be carried out as though the subject were in hypnosis.

CONSCIOUS MIND: Normal everyday state of consciousness, normally associated with the Beta brain wave pattern.

SUB-CONSCIOUS MIND: Awareness beneath the normal conscious level. This level is normally associated with the Alpha brain wave pattern.

UNCONSCIOUS MIND: Unconscious aspect of the mind such as breathing, sleeping, and other functions necessary for life. This is usually associated with the Theta and Delta brain wave patterns.

CATALEPSY: State of muscle rigidity induced in a subject by an operator.

SOMNAMBULISM: State of deep lethargy induced by suggestion, with amnesia upon awakening.

LEVELS OF HYPNOSIS:

Although many authors use from three to fifty levels to describe various stages of hypnosis, for the purposes of this course we will use twelve. They can be categorized into segments of three but will give the student a finer insight into technique development than could otherwise be achieved. The levels that we will deal with are as follows:

1.) GENERAL RELAXATION: The subject's body is observed to be relaxed all over. The subject is instructed to be as comfortable as possible and to place his feet flat on the floor and hands are to remain in the lap unclasped.

2.) EYE CLOSURE: Subject's eyes will tend to close when proper suggestions are given. As they close, the subject will notice some occasional light beneath the eyelids and the operator will notice that the lids tend to flutter somewhat.

3.) MUSCLE RIGIDITY: Subject is given suggestions to make his arm stiff and rigid, so much that neither the hypnotist nor the subject himself can bend it. He is told to try to bend it but cannot.

4.) MOTOR MOVEMENTS: Subject's arms are put into a rotating motion and he is instructed to stop them after being told that he cannot.

5.) PARTIAL ANESTHESIA: After appropriate suggestions have been given, the subject's hand is pinched and he responds with some indication of discomfort.

6.) TOTAL ANESTHESIA: The subject's hand is pinched with no response of discomfort.

7.) PARTIAL RESPONSE TO POST-HYPNOTIC SUGGESTION: When brought out of the hypnotic state, subject will only partially respond to appropriate suggestions given while in the hypnotic state.

8.) TOTAL RESPONSE TO POST-HYPNOTIC SUGGESTIONS: Subject responds completely to appropriate suggestions given while in the hypnotic state after awakening.

9.) POSITIVE VISUAL HALLUCINATIONS: When brought out of the hypnotic state, subject will see objects that are not present in reality.

10.) NEGATIVE VISUAL HALLUCINATIONS: Subject will not see objects that are, in fact, present.

11.) PARTIAL AMNESIA UPON AWAKENING: Subject will have hazy memory of what has happened while in hypnosis.

12.) TOTAL AMNESIA UPON AWAKENING: Subject will have no memory of what has taken place while in the hypnotic state.

13.) HYPNOTIC COMA (Ultra-Depth): Little is known about this state of mind. Subjects entering this level do not respond normally to suggestions as they would in the normal range of hypnosis.

CHAPTER TWO

PRELIMINARY TESTING

Before the induction of hypnosis it is often a good idea to administer various tests to determine what type of response can be expected. This will enable the therapist to make an adjustment in technique or methodology in order to achieve maximum efficiency. Some people can be induced into hypnosis within a few minutes while others require several sessions before an adequate level of hypnosis can be achieved.

You will discover that about 20% of the general population is highly suggestible and make excellent subjects, about 60% are average subjects and can be developed into good subjects, about 15% are very difficult and will require the therapist's greatest skill to achieve good results and there are about 5% that the therapist cannot hypnotize. You will find that those who suffer from mental retardation and others with extreme ego problems generally cannot be hypnotized.

PRELIMINARY TESTING TECHNIQUES:

1.) HEAVY AND LIGHT ARM: The subject is told to extend his arms out directly in front of him and close his eyes. He is then told to imagine that a heavy book is in his right hand and that a gas-filled balloon is tied to his left wrist. He is told that his right arm is getting heavier and is going down while his left arm is floating upward toward the ceiling. You may determine his degree of suggestibility by his response.

2.) HAND CLASP TEST: The subject is instructed to clasp his hands together with the fingers interlaced. He is then told to stare at one thumb or to close his eyes, whichever is desired. The subject is then told to imagine that a rope has been tied around his hands and to clasp his hands together tighter. These suggestions are repeated and the subject is told that the rope has bound his Hands so tightly together that he cannot pull them apart no matter how hard he tries.

3.) FALLNG BACKWARD: The subject is told to close his eyes and roll the eyeballs upward as though he were looking at a spot above his head. The hypnotist then tells the subject that he is being pulled backward into a soft pile of hay or onto several pillows and that he cannot resist the backward pull. The suggestions are repeated and the hypnotist stands behind the subject to prevent the fall when the subject responds.

4.) PROGRESSIVE RELAXATION: In this test the subject is given fractional relaxation suggestions to test for his ability to follow instructions in releasing tension.

5.) SWINGING PENDULUM: In this test a subject is given a pendulum on a string and is instructed to hold it perfectly still over a given point. The hypnotist then gives instructions that it is being pulled to the right and then to the left. The pendulum will be noted to begin swinging in accordance with the suggestions given.

CHAPTER THREE

EXTERNAL FACTORS

When attempting to induce hypnosis, there are many influencing factors that have to be dealt with before a satisfactory depth can be obtained. If any of these factors are ignored, far less success will be obtained than if prudent observation is made and one's techniques are adjusted accordingly.

1.) PERSONALITY CLASHES: If it is determined that there is a personality clash, it is far better to immediately transfer the subject to an associate therapist than to attempt to work with the subject under such a handicap.

2.) PROFESSIONALISM: It is essential that the hypnotherapist project himself to the subject as a professional person and conduct himself accordingly during the entire duration for therapy purposes. Much greater results will be obtained if this rule is followed.

3.) OFFICE ARRANGEMENT: The hypnotherapist's office should be large enough to imply success. The furniture should be at least moderately expensive and comfortable, reflecting good taste.

4.) SECRETARY AND/OR RECEPTIONIST: There should never be an answering machine or answering service handling the phone during office hours. A carefully selected person should greet all clients and handle appointments and collections. The office personnel are very crucial to success, as they are the first to begin conditioning your clients to hypnosis.

5.) TELEPHONE VOICE: The person answering the phone is the first contact that most clients have with hypnosis. The manner in which the phone is handled can make or break the entire business. Clients who are poorly handled will not come in or will make poor subjects when they do come in.

HYPNOTIC TECHNIQUES

There are some hypnotherapists who prefer to do all of their work on a one-to-one basis with the subject. There are others who prefer to start out on a personal basis and then switch to tapes for the follow-up sessions. There are pros and cons on both sides and we will discuss each separately.

1.) PERSONAL INDUCTIONS: It is essential that all induction be performed personally at first. This establishes rapport with the subject and gives the hypnotist first hand experience of what responses he can expect from a given client. However, when working with female clients, it is best for a male therapist to have a second person present during the work.

2.) TAPE INDUCTION AND FOLLOW-UP: Once a subject has been successfully induced into the hypnotic state and worked with for problem clarification on a personal basis, he can then be transferred to tapes by appropriate suggestions. This releases the hypnotherapist to work with other subjects and also afford the subject total privacy.

Some people are not as responsive to tapes as to personal work and the hypnotherapist must decide which method to use to obtain maximum benefit to the client.

3.) TAILOR THE TECHNIQUES TO THE PERSON: The client must be observed carefully as to what he or she is responding to and the operators techniques must be administered appropriately for the overall benefit to the client.

There will be times when the operator's entire capability is exhausted and the client has derived no benefit. This is the time to switch to another therapist or make recommendations for a different type of therapy.

Techniques should be varied to insure maximum depth for the work to be done. In some cases the light to medium state is adequate whereas in other cases the subject must be worked into as deep a hypnotic state as possible.

EMOTIONAL FACTORS

1.) AN UPSET SUBJECT: When a subject is upset from external factors, he can generally be helped effectively. However, if the upset is with the operator, it is quite a different story.

A hypnotized subject will sometimes spontaneously relive a past trauma and exhibit hysterical behavior. This usually terrifies the novice. However, by remaining calm and in firm control, the therapist can assist the subject in releasing the trauma associated with the event and proceed forward with the general work at hand.

If the hypnotist must suddenly leave his subject due to an emergency, it is imperative that he awaken the subject normally or switch therapist smoothly and normally and not upset the subject with his own personal drama.

2.) ADMIRATION OF THE HYPNOTHERAPIST AS A POSITIVE AND NEGATIVE EFFECT:

The subject's respect and admiration of the therapist is a positive factor in obtaining good results for any given subject. However, this admiration must not be converted into an "ego trip" by the hypnotherapist or the overall good that could have been accomplished will be destroyed.

3.) DIPLOMAS: Any certificate, diploma or other credential that has been honestly earned can be displayed appropriately. Bought diplomas or "mail order degrees" which are obtained merely to impress someone will be of little value to your clients and become a discredit to you and your profession when they are dis-covered by other practitioners.

ADDITIONAL FACTORS:

1.) Your hourly rate must be appropriate for the service rendered and must be acceptable to the average client. Normally hypnosis clients do not stay for great lengths of time so their overall cost is generally lower than for other means of therapy.

2.) SPOUSAL OPPOSITION OR SUPPORT: The emotional support of your client on the home front makes all the difference in the world. If he does not have home support he may as well be declined for therapy because a negative spouse can undo all the good that a therapist can do. However, if the spouse is either supportive or simply is not aware of your client's activities, much good may be accomplished.

3.) RESULTS: If your client does not notice immediate results, he will oftentimes decide that he has not been hypnotized and stop coming. A complete explanation of what hypnosis is and what it is not should be covered by the hypnotist prior to beginning work.

A subject should understand that hypnosis is a normal state of mind that he enters at least twice per day when he is waking up in the morning and just before he falls asleep at night. This will help him to understand that hypnosis is natural state of mind and that he will be consciously aware during the entire process.

CHAPTER FOUR

INDUCTION TECHNIQUES

The actual induction monologue will be given in a later segment. The technique discussed here can be utilized with the general monologue and varied in accordance with the hypnotist's personality and general preferences.

1.) SWINGING PENDULUM: With this method the subject is instructed to watch a pendulum, either actual or imaginary, and listen to the hypnotist. The hypnotist then goes through the general monologue referring to the pendulum swinging as progressively increasing relaxation on the part of the subject.

When adequate relaxation has been achieved, the suggestions are then given for heavy eyelids or further motor movement testing.

2.) EYE FIXATION ON A STATIONERY POINT: The subject is instructed to locate a point above his normal eye level and to look only at that point. The monologue is then gone through in the usual manner.

3.) COUNTING AND/OR PYRAMIDING: The operator begins counting backward and instructs the subject that with each count he becomes twice as relaxed as before.

4.) VISUALIZATION: The subject can be taken on imaginary walks through parks or on trips that will enhance the depth of hypnosis considerably. With subjects whose imagination is poor, this method is not recommended.

5.) STROKING: As the general monologue is given the subject's arm or hand is stoked by the operator. The shoulder is pressed down in accordance with the appropriate suggestions given at that time.

6.) SILENT: This technique is effective on those who have seen how people look and react under hypnosis. It is accomplished by having the subject close his eyes and then utilizing the stroking method but with no use of verbalization. The process of variously emphasizing pressure on the arms and shoulders will convey to the subject what is expected of him and he will respond and enter into a deep state of hypnosis without a word having been spoken.

7.) PROGRESSIVE RELAXATION: This is a simple method of beginning with the subject's toes and gradually relaxing his entire body upward to the top of his head or moving from the head to the toes.

8.) SLEEP-AWAKE-SLEEP: The subject is relaxed into as deep a state as possible and then told that he will be awakened. When he is awakened, he will then be told that he will now be returned to the deeply relaxing state that he was in before he was awakened but he will go much deeper than before. This can be repeated many times and is good for resistant types who would find this method more dramatically convincing than other methods.

Many of these techniques are interchangeable and can be used in any combination of methods to induce hypnosis in a given subject. This is why it is important to know many different methods and techniques of hypnosis and how the subject should react to them. The greater knowledge of methodology that you have, the greater will be your power to effectively lead your subject into hypnosis.

DEEPENING PROCESS:

In the event that the subject has not reached an adequate depth for work to begin, there are deepening methods that can be utilized for greater depth enhancement.

1.) VISUAL ASSISTANCE: When the subject is as relaxed as possible, he is told to open his eyes and look into the eyes of the hypnotist. He is further told that when he does he will turn loose and allow himself to enter into a deeper and deeper state of relaxation. This can be repeated several times. Each time that the subject opens and closes his eyes deepening suggestions can be reinforced.

2.) STROKING: The subject can be told that as the operator strokes his hand or arm he will become more and more relaxed and more receptive to positive constructive suggestions. The subject's hand or arm can then be stroked or lifted and allowed to fall into his lap.

3.) HAND AND ARM LEVITATION: The subject is told that as he relaxes more and more his right arm is becoming lighter and that he will not restrain it from floating upward toward the ceiling. The suggestion is repeated several times until movement is obtained. The subject is then told that upon the snap of a finger his arm will fall and he will be even more relaxed than before.

4.) PEACEFUL REGRESSIVE TECHNIQUE: The method allows communication between the subject and the operator without disturbing the subject's depth that has already been achieved. He is told to remember a beautiful, peaceful scene that he has visited in the past and to recall it just as though he were actually there and to feel the relaxation that he felt when he was there.

5.) TENSION-RELAXATION ALTERNATION: When the subject is as relaxed as possible, have him tighten his stomach muscles tighter and tighter and to release them upon the count of three. By doing this a few times and giving additional deepening suggestions a greater level of relaxation can be obtained.

6.) ELEVATOR: The subject is told to visualize himself entering an elevator on the tenth floor. He is instructed to imagine watching the floor numbers as he descends. He is to become aware that he is becoming more and more relaxed as the elevator approaches the first bottom level.

There are many hundreds of methods for induction and deepening and by practicing these, you will have the basis for development of other methods of your own that will work for you better than anything you may read about. However, we all have to start with concrete beginnings before we can fly, so these basic methods are given for your initial experiences.

CHAPTER FIVE

INDUCTION MONOLOGUES

The following monologue can be paraphrased or used exactly as presented to induce a hypnotic state in someone else or for self-hypnosis purposes. It is suggested that you study each monologue several times and that you become totally familiar with the material before administering the described technique(s) to your subject. Please note that it is far better for your new subject that he is not aware that he is your first attempt at hypnosis. Maintain an air of professionalism and skill. Even physicians are nervous when first learning new medical techniques. It is suggested that a tape recording of the monologue be made to discover where improvements can be made in the smoothness of vocal presentation and transition from one level of hypnosis to the next.

You will also notice that within the monologue are tests to allow you to determine when your client has entered each respective level of hypnosis. Once you have mastered the induction and are familiar with the depth that your client has attained and can recognize it without testing, then you can discard all of the tests in the monologue and proceed directly with deepening and other procedures designed for your client's therapy.

INDUCTION MONOLOGUE:

"Allow your entire body to relax and begin turning loose all of your muscles. Relax each and every nerve, each and every muscle. Relax each and every nerve, each and every muscle. Let all of the muscles of your body relax completely...completely... totally...and completely.

All of the muscles of your entire body are relaxing now completely…totally…and you are allowing yourself to enter into a deep…wonderful…relaxing…state of mind. A wonderful…deep…deep…state of peace and rest. All of the muscles of your body are now becoming completely and totally rested and your mind is becoming receptive to all suggestions that will bring improvement into your life.

All of the muscles of your body are now relaxed completely and your eyelids are now becoming very heavy…very heavy…they get heavier and heavier…h-e-a-v-i-e-r and h-e-a-v-i-e-r and you go d-e-e-p-e-r and d-e-e-p-e-r into rest and relaxation… You feel as though you could enter a wonderful deep state of rest and peace…d-e-e-p…d-e-e-p…rest and relaxation. Your eyelids are getting h-e-a-v-i-e-r and h-e-a-v-i-e-r and you cannot hold them open any longer.

your eyes are closing now and as they close, you feel yourself going d-e-e-p-e-r and d-e-e-p-e-r into peace and rest. (Throughout this procedure observe the subject's eyes for heaviness and when the eyes are closed, notice any flutter of the eyelids.)

You are now in a deep, wonderful state of peace and rest and when you try to open your eyes, you will notice that you have relaxed them so much that they feel as though they weigh a ton. The harder you try to open your eyes the h-e-a-v-i-e-r they will feel. Go ahead now, try to open your eyes…you cannot…no matter how hard you try…you cannot open your eyes… Now stop trying and allow yourself to enter an even deeper state of peace and rest.

You are now in a deep wonderful state of rest and peace and when I raise your arm you will go even deeper into this wonderful rest. I will pick up your arm now and you will go d-e-e-p-e-r and ever d-e-e-p-e-r into rest and relaxation. Now as I pick up your arm I am going to hold it straight out in front of you. As I do so I want you to put all of the residual tension in this arm. Make your arm stiff and rigid! Make this arm so stiff and rigid that it feels like a bar of steel and just as strong. Take all remaining tension in your body and put it into this arm! Make it so stiff and rigid that I cannot bend it! (Try to bend his arm and if it stiff enough, proceed.)

I cannot bend your arm and neither can you! Neither you nor I can bend your arm, try to bend it now! Try to bend your arm and you find that you cannot! Try again! Try again!… Now, totally release all tension in your arm and allow it to fall into your lap completely loose and relaxed. Let your arm fall into your lap now totally l-o-o-s-e, l-i-m-p and r-e-l-a-x-e-d. (You have now achieved the third stage of hypnosis with your subject; muscle rigidity.)

Now you are more relaxed that you have ever known before. However, in order to allow you to go even deeper, I am going to begin to rotate your hands. (Pick up subject's hands and begin to rotate his hands.) This now becomes a totally automatic motion and they will continue to rotate even when I remove my hands. When I remove my hands, your hands will continue to rotate easily and effortlessly. Now as I remove my hands, your hands will rotate easier and easier. (Remove your hands and notice that the subject's hands are rotating easily.)

Now, when I snap my fingers, your hands will rotate faster and faster. When I snap my fingers again, you hands will rotate in the opposite direction and even faster. (Snap your fingers and notice the change in direction.) Now, you are unable to stop your hands from rotating. The harder you try the faster they rotate. Try to stop them now! Try to stop them now!...(Observe the subject's futile efforts to stop his hands from rotating.) Now when I snap my fingers, your hands will fall into your lap loose and relaxed and you will go even d-e-e-p-e-r and d-e-e-p-e-r into total rest and relaxation…deeper and deeper with every breath you take. (Snap finger and observe total collapse of tension in the arms. This is attainment of the fourth stage of hypnosis.)

At this point, continue with the general monologue and relax your subject further and proceed with the test for the fifth stage of hypnosis.

As I stroke your hand it will become very numb..very numb..v-e-r-y...n-u-m-b… your hand is becoming so numb that you can feel no pain or discomfort in this hand as all…as it has become so very numb…so…very…numb…no pain or discomfort at all.

Your monologue has carried your subject to the eighth stage of hypnosis. With additional suggestions and deepening techniques, you can carry him on to the twelfth stage, or somnambulism. Testing your subject each step of the way is not necessary when you become familiar with your profession and begin to recognize when a subject is at the proper degree of receptivity to begin positive suggestions for therapy.

SELF-HYPNOSIS INDUCTION MONOLOGUE:

The major difference between hypnotizing others and oneself is that all of the suggestions have to be given to our own mind and, consequently, the monologue must be reworded accordingly. In addition, we must take a slightly different approach when hypnotizing oneself because we must be alert and yet receptive and relaxed at the same time. We have to administer positive suggestions to ourself and be able to accept them at the subconscious level on a concurrent basis.

The best method for self-hypnosis is somewhat more passive and less challenging that the other hypnotic methods as the following monologue will illustrate:

"I am now very comfortable and relaxed. As I begin counting backward from ten to zero, I will become even more comfortable and relaxed than ever before. I will become so much more comfortable that all of the muscles of my body will be totally limp, loose and relaxed. By the count of zero, all of the muscles of my body will be completely loose, limp and relaxed...loose...limp...and relaxed.

Ten, I am now comfortable and relaxed but will become even more relaxed as I count further down. Nine, I am now allowing all of the muscles of my right leg to

Now as I press your hand, if you feel any discomfort at all, raise yo

opposite hand. Now as I press your numb hand, you will fe

(Reinforce the signal suggestions for the raising of the finger of t

pinch the numbed hand. If slight discomfort is felt he is in the fifth

if no discomfort is felt, he is in the sixth stage of hypnosis. When

remove all numbing suggestions and move on the next level.)

You are now very deeply relaxed. Soon, however, you will b

wonderful and refreshed. You will feel as though you have had a ver

rested and refreshed. When you are awakened, you will notice that

sitting on some glue on the floor and you will notice that you cannot

the very spot it is in right now. The harder you try to move your foo

will stick it to the floor. Even though you will be wide awake, yo

move your foot until I say the word "RELEASE". When I say the wor

that time your foot will then be free of the glue. One, you are very re

are going to awaken. Two, the next time will be even deeper than

gradually awakening now…Three, coming back to the everyday v

almost awake now…five, wide awake…wide awake, fully alert and aw

At this point you will notice the subject's reaction to your sugge

under hypnosis. If your subject cannot move his foot, then you have a

stage of hypnosis. If your subject hesitates but eventually moves

seventh stage has been achieved. Even if your subject did not respond

hypnotic level, always give the releasing word anyway as a safety precau become completely loose, limp and relaxed…completely loose, limp and relaxed. Eight…I am now allowing all of the muscles of my left leg to become completely loose, limp and relaxed, completely loose…limp…and relaxed. Both of my legs are now completely loose, limp and relaxed. They feel so relaxed that if I tried to move them they would be heavy as though they were made of lumps of lead, heavy lumps of lead…heavy…heavy…lumps of lead. Seven…now this wonderful feeling of rest and relaxation is spreading into my lower back and abdominal area. My entire lower back and abdomen are now becoming totally loose…limp…and relaxed. Six..I am now allowing this relaxation to spread into my stomach and chest…this relaxation is now spreading into my stomach and chest…my stomach and chest are now feeling so wonderfully relaxed…so loose…limp…and relaxed…totally…deeply and completely relaxed. Five…my right shoulder…right arm…right forearm…right wrist…hand and fingers are now totally lose…limp…and relaxed…loose…limp and relaxed. Four…my left shoulder…left arm…left forearm…left wrist, hand and fingers are also totally loose…limp…and relaxed…loose…limp…and relaxed. Three…this relaxation is now flowing up into my neck…into the back of my neck and head…flowing up into my scalp…flowing into my scalp and now down over my forehead…down over my eyes, nose, lips, mouth, tongue, throat and jaw muscles… Two…my entire body is now totally and completely relaxed…totally and completely relaxed…loose…limp…and relaxed… One…I am now completely…totally…and in every way…relaxed…loose…limp…and relaxed… I am now completely loose…limp…and relaxed…and am now totally receptive to positive, constructive ideas and suggestions that will bring improvement into my life. I am now ready for suggestions that will benefit me in every way and become part of my life. All suggestions that I now hear or say to myself will take total effect upon me and I will accept them totally and and completely and I will act upon them in a positive…constructive…manner.

(The suggestions are now to be given either mentally or by means of pre-recorded tapes that can be turned on or off by yourself without disturbing your deep rest whatsoever.)

The suggestions that I have now heard are for my benefit and self-improvement. I accept them as truth and fact and will act upon them, as they are now my ideas… beliefs…and goals. These ideas are now my ideas…they are a part of my entire being and will be reflected externally in my life as well as internally. I will act upon them in a positive manner and utilize them to bring improvement into my life.

Soon, I will bring myself out of this wonderful, beneficial state of mind. When I awaken, I will feel good, refreshed and invigorated… I will feel good, positive and enthusiastic. I have faith and confidence in the suggestions that I have now been given and know that they are for my self-improvement and will work for me. Each time that I practice my self-hypnosis techniques, I become better and better…each time that I practice I become better and more proficient than the previous times. The suggestions that my mind are presented with are designed for my benefit and improvement for I will accept no suggestions that are contrary to my betterment and the betterment of those around me.

Upon the count if five, I will awaken fully alert, refreshed, invigorated and optimistic. One…I am now very relaxed but will return to the normal waking state of mind. Two…I am awakening now…slowly…each time I practice my self-hypnosis, I

Now as I press your hand, if you feel any discomfort at all, raise your index finger of the opposite hand. Now as I press your numb hand, you will feel no discomfort all. (Reinforce the signal suggestions for the raising of the finger of the opposite hand and pinch the numbed hand. If slight discomfort is felt he is in the fifth sage of hypnosis and if no discomfort is felt, he is in the sixth stage of hypnosis. When this test is finished, remove all numbing suggestions and move on the next level.)

You are now very deeply relaxed. Soon, however, you will be awakened and feel wonderful and refreshed. You will feel as though you have had a very refreshing nap - so rested and refreshed. When you are awakened, you will notice that your foot has been sitting on some glue on the floor and you will notice that you cannot move your foot from the very spot it is in right now. The harder you try to move your foot the tighter the glue will stick it to the floor. Even though you will be wide awake, you will be unable to move your foot until I say the word "RELEASE". When I say the word "RELEASE"…at that time your foot will then be free of the glue. One, you are very relaxed now but you are going to awaken. Two, the next time will be even deeper than this time. You are gradually awakening now…Three, coming back to the everyday waking state, four, almost awake now…five, wide awake…wide awake, fully alert and awake!

At this point you will notice the subject's reaction to your suggestions given while under hypnosis. If your subject cannot move his foot, then you have achieved the eighth stage of hypnosis. If your subject hesitates but eventually moves the foot, then the seventh stage has been achieved. Even if your subject did not respond at all to the post-hypnotic level, always give the releasing word anyway as a safety precaution.

Your monologue has carried your subject to the eighth stage of hypnosis. With additional suggestions and deepening techniques, you can carry him on to the twelfth stage, or somnambulism. Testing your subject each step of the way is not necessary when you become familiar with your profession and begin to recognize when a subject is in the proper degree of receptivity to begin positive suggestions for therapy.

SELF-HYPNOSIS INDUCTION MONOLOGUE:

The major difference between hypnotizing others and oneself is that all of the suggestions have to be given to our own mind and, consequently, the monologue must be reworded accordingly. In addition, we must take a slightly different approach when hypnotizing oneself because we must be alert and yet receptive and relaxed at the same time. We have to administer positive suggestions to ourself and be able to accept them at the subconscious level on a concurrent basis.

The best method for self-hypnosis is somewhat more passive and less challenging that the other hypnotic methods as the following monologue will illustrate:

"I am now very comfortable and relaxed. As I begin counting backward from ten to zero, I will become even more comfortable and relaxed than ever before. I will become so much more comfortable that all of the muscles of my body will be totally limp, loose and relaxed. By the count of zero, all of the muscles of my body will be completely loose, limp and relaxed...loose...limp...and relaxed.

Ten, I am now comfortable and relaxed but will become even more relaxed as I count further down. Nine, I am now allowing all of the muscles of my right leg to become completely loose, limp and relaxed…completely loose, limp and relaxed. Eight…I am now allowing all of the muscles of my left leg to become completely loose, limp and relaxed, completely loose…limp…and relaxed. Both of my legs are now completely loose, limp and relaxed. They feel so relaxed that if I tried to move them they would be heavy as though they were made of lumps of lead, heavy lumps of lead…heavy…heavy…lumps of lead. Seven…now this wonderful feeling of rest and relaxation is spreading into my lower back and abdominal area. My entire lower back and abdomen are now becoming totally loose…limp…and relaxed. Six..I am now allowing this relaxation to spread into my stomach and chest…this relaxation is now spreading into my stomach and chest…my stomach and chest are now feeling so wonderfully relaxed…so loose…limp…and relaxed…totally…deeply and completely relaxed. Five…my right shoulder…right arm…right forearm…right wrist…hand and fingers are now totally lose…limp…and relaxed…loose…limp and relaxed. Four…my left shoulder…left arm…left forearm…left wrist, hand and fingers are also totally loose…limp…and relaxed…loose…limp…and relaxed. Three…this relaxation is now flowing up into my neck…into the back of my neck and head…flowing up into my scalp…flowing into my scalp and now down over my forehead…down over my eyes, nose, lips, mouth, tongue, throat and jaw muscles… Two…my entire body is now totally and completely relaxed…totally and completely relaxed…loose…limp…and relaxed… One…I am now completely…totally…and in every way…relaxed…loose…limp…and relaxed… I am now completely loose…limp…and relaxed…and am now totally receptive to positive, constructive ideas and suggestions that will bring improvement into my life. I am now ready for suggestions that will benefit me in every way and become part of my life. All suggestions that I now hear or say to myself will take total effect upon me and I will accept them totally and and completely and I will act upon them in a positive…constructive…manner.

(The suggestions are now to be given either mentally or by means of pre-recorded tapes that can be turned on or off by yourself without disturbing your deep rest whatsoever.)

The suggestions that I have now heard are for my benefit and self-improvement. I accept them as truth and fact and will act upon them, as they are now my ideas…beliefs…and goals. These ideas are now my ideas…they are a part of my entire being and will be reflected externally in my life as well as internally. I will act upon them in a positive manner and utilize them to bring improvement into my life.

Soon, I will bring myself out of this wonderful, beneficial state of mind. When I awaken, I will feel good, refreshed and invigorated… I will feel good, positive and enthusiastic. I have faith and confidence in the suggestions that I have now been given and know that they are for my self-improvement and will work for me. Each time that I practice my self-hypnosis techniques, I become better and better…each time that I practice I become better and more proficient than the previous times. The suggestions that my mind are presented with are designed for my benefit and improvement for I will accept no suggestions that are contrary to my betterment and the betterment of those around me.

Upon the count if five, I will awaken fully alert, refreshed, invigorated and optimistic. One…I am now very relaxed but will return to the normal waking state of mind. Two…I am awakening now…slowly…each time I practice my self-hypnosis, I gain more positive control of myself and am able to influence my life in a more positive manner… Three…I am coming awake now…I am waking up… Four…almost awake… Five…I am now wide awake, fully alert and awake"

CHAPTER # 6

ADDITIONAL FACTORS

The monologue just presented can be altered to fit the personality of the hypnotist. The challenges can be deleted and both monologues can be made into more passive phrases or the self-hypnosis can be made more into a testing method. They are fundamental building blocks and are to be used as such in order to build experience and confidence in dealing with the hypnotic state in both the auto and heterohypnotic situations.

AWAKENING METHODS:

1.) SIMPLE COUNTING: Most of the time simply counting from one to five will suffice. However, in other cases, more stringent methods are necessary.

2.) LONG COUNT: At times, the subject is so relaxed and contented with the state of rest that he has come to enjoy that here is some reluctance to come out of it. In this event, a longer count from one to ten or higher may be necessary. With each number the idea of future sessions being even more enjoyable is an asset.

3.) RESISTANT SUBJECT METHODS:

A.) DIRECT COMAND APPROACH: This is done by deepening the state and then having the subject perform activities by suggestions and then giving the command to awaken in an authoritative manner.

B.) PERMISSIVE APPROACH: This method simply allows the subject to awaken whenever he wants to do so. Sometimes there will be a five or ten minute wait, depending upon the subject.

C.) SLEEP APPROACH: Simply let the subject goes to sleep and he will awaken normally and be refreshed and alert.

4.) WHEN ALL OTHER METHODS FAIL:

    A.) WALK SUBJECT AROUND: Exercise through suggestion.

    B.) LET SUBJECT SLEEP IT OFF OVERNIGHT: Or possibly for several hours.

    C.) CALL IN MEDICAL ASSISTANCE: Only as a last resort.

There is no case on record of anyone not awakening from hypnosis. There are, however, numerous cases of hypnotists who panicked and had the subject rushed to a hospital for a doctor to awaken him. Had the hypnotist been properly trained and skilled, those events would never have transpired.

PROBLEM SOLVING METHODS:

The subconscious mind will give us back what we put into it and with interest. If we program it to give us an answer to our problem, it will do just that. If we program it to keep us out of problematic situations, we will find that our lives will become more trouble free and we can live the type of life we want by controlling our mind.

1.) TV TECHNIQUE: When in self-hypnosis or administering hypnosis to someone else, the suggestions can be given that the subject can visualize a TV and the solution to a particular problem will appear on the screen.

2.) DELAYED ANSWER TECHNIQUE: The mind can be programmed to dwell on the solution rather than the problem and that the answer will pop into the mind within a short time after concentration on the solution begins, usually within a day to a week.

3.) MIRROR TECHNIQUE: The subject is told to visualize himself looking into a mirror and to see the image being free of the particular problem at hand. Once this is seen clearly, he is told to then blend the image in the mirror with himself and to believe that his problem has been solved.

Additionally, while looking at the image in the mirror, the subject can attribute to the image any good quality that he desires to be incorporated into his own personality and then blend with that image and it will become a reality

MECHANICAL AIDS:

1.) PENDULUM: This device is useful in the beginning for the subject to watch as the hypnotist swings it back and forth. It tires the eyes rather quickly and thus enhances eye closure.

2.) LIGHT: Having a small penlight to shine on the closed eyelids of your subject can be most helpful and effective as the light is at first intense and then gradually pulled away, simulating a visual dimming and effectively implementing a deepening state of relaxation.

3.) METRONOME: This device is extremely useful, not only in the initial induction and deepening process but also in the making of tapes. The constant ticking is excellent background and can be varied in accordance with the situation at hand.

4.) REVOLVING SPIRALS: These can be purchased from any hypnotic supply houses and are good induction aids.

5.) FIXATION POINTS: This is simply a pin, dot or other noticeable spot that the subject is told to fix his attention for the induction. As he continues to gaze at the spot, his eyes become very tired and his induction process is speeded up.

6.) BRAIN-WAVE SYCHRONIZER: This is an electronic machine that will match the alpha-wave cycle of the brain with a pulsating light against the back of the eyelids and is extremely effective in the induction process.

7.) BACKGROUND SOUND: A consistent sound of water, ocean waves, fire, or a constant hum or hiss if often very helpful. It also tends to absorb or block out other intrusive noises.

8.) MUSIC BACKGROUND: This is tricky. Some music may evoke unwanted reactions within your subject. This is best avoided unless you are experienced with just what your subject enjoys and reacts to in a favorable manner.

9.) SPECIAL CHAIRS: Vibrating, reclining chairs or those that have heat and rollers are very good for the hypnotic induction and continued follow-up therapy. Straight chairs are fine for short duration personal induction and some limited therapy.

10.) AUTOMATED HYPNOSIS: Hypnosis administered through earphones into private client workrooms by means of automatic tape changers can be effectively accomplished through personal work followed by tape conditioning. This method is usually done in larger medically oriented clinics.

DANGERS OF HYPNOSIS:

1.) FORGETTING TO REMOVE THE SUGGESTIONS: All suggestions given to a subject for testing purposes should be removed prior to bringing them out of trance state. All arm stiffness, anesthesia, deep relaxation or other suggestions should be cleared before the waking process begins.

2.) CONFABULATION: This is when a female subject fantasizes sexual advances by the hypnotherapist. This is best prevented by having another person present during therapy or by taping the entire session from the time that the lady enters the presence of the therapist until the session is concluded and she has left the area.

3.) HYSTERIA: There are times when the hypnotized subject will invariably regress back to some previous traumatic event and exhibit hysterical behavior. Calmness and positive control on the part of the hypnotist can assist the subject to release the emotional content of the trauma and to relax and proceed with the session at hand.

4.) PHYSICAL DANGERS:

>A.) HEART PATIENTS: Always work with heart patients through their doctor's written authorization. In the event of your client having a heart attack during a session, this will ensure your protection from legal action and liability.

>B.) EMOTIONALLY DISTURBED INDIVIDUALS: These people are best dealt with by medical personnel. Emotionally disturbed individuals must be immediately referred to a physician.

5.) OTHER DANGERS:

   A.) TRANSFERENCE: This can generally be avoided by maintaining a professional therapeutic relationship with your client and by having a co-worker present or nearby at the time the therapy is administered. Also, tape sessions are favorable for prevention of this problem because the therapist is not physically present with the client.

   B.) MALPRACTICE: This usually happens when the hypnotist over-steps his bounds and attempts to practice medicine or psychology or encroaches into other areas where he is not certified nor licensed.

   C.) PRODUCING SUDDEN CHANGES IN THE SUBJECT: Avoid sudden changes within the subject's mind. Maintain consistency and make the changes gradual. Sudden emotional changes can cause extreme reactions.

CONCLUSION

This material is a synopsis of what taught at the facilities of the American Institute for Human Development in Norcross, Georgia for many years. It represents a basic understanding of the fundamentals of hypnosis and is not to be construed as anything more than fundamental.

The student is encouraged to attend other classes held by professionals in the field if he or she desires to actually engage in professional hypnotherapy. Hypnotherapy is an advanced state of the art that dedicated people will pursue if their interest is deep enough to help their fellow man

ADDITIONAL BOOKS

And other information by:

Larry M. McDaniel

Faith, the Secret to Love and Success

Change Your Words – change your Life

Detection of Deception With Handwriting Analysis

Handwriting Analysis and Employment Screening

Pemlar Profile – Pre-Employment Test

Hypnosis Basic – 101 – The Beginning of Knowledge

Hypnosis 201 – Advanced Techniques

Hypnosis 301 – Hypnotherapy

Hypnosis 401 – Hypnoanalysis – Psychoanalysis

Hypnosis to Hypnoanalysis – The Ultimate Therapy

Homosexuality and Pedophilia – The Cause and Cure

Post Traumatic Stress Disorder – The Cause and Cure

Lulu.com/spotlight/LarryMcDaniel

LarryMcDaniel12@GMail.com

44

www.ingramcontent.com/pod-product-compliance
Lightning Source LLC
Chambersburg PA
CBHW081020170526
45158CB00010B/3110